# BASTI
## The Ring-Tailed Lemur

BY

**Anthony D. Marshall**

A WINDSWEPT BOOK
Windswept House Publishers
Mt. Desert, Maine 04660

Library of Congress Catalog Number 87-051167
ISBN 0-932433-37-5

Printed in the United States of America
For the Publisher by
Downeast Graphics and Printing, Inc.
Ellsworth, Maine

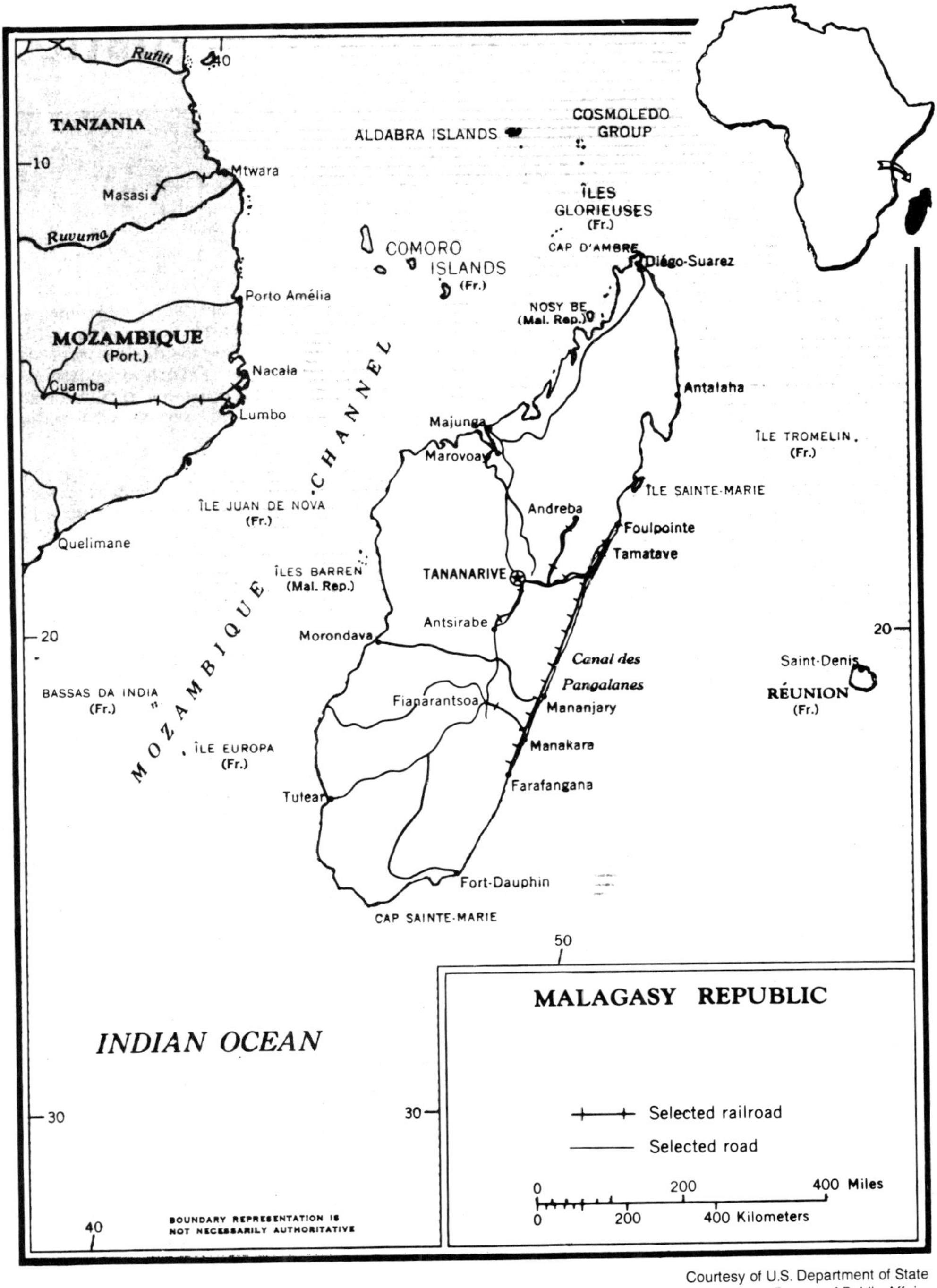


Courtesy of U.S. Department of State
Bureau of Public Affairs

# BASTI
## The Ring-Tailed Lemur

. . . . "the free hours I
needed so badly for
writing — outdoors in the
garden with Sebastian" . . . .

*In a letter from*
*Charles A. Lindbergh*
*November 13, 1970*

## Chapter One

This is a true story.

My name is Sebastian. My friends call me Basti. But this isn't the name I was born with. I was born without a name in the summer of 1970 in Madagascar. The great island country of Madagascar is 250 miles off the coast of Africa in the Indian Ocean. Unlike some places in Africa where Man lived a million years ago, no man set foot on Madagascar until 200 years after the birth of Christ. However, my ancestors have been on Madagascar for longer than Man in Africa.

I am a lemur. In Latin I am called a lemur catta. The Malagasy call me a maki. In English I am a ring-tailed lemur. The French call us *singes*, or monkeys, which we are not! You may call me Basti.

Let me tell you about lemurs. When life was created on this planet, there eventually developed monkeys and apes, but we lemurs went our own way on the Tree of Life 40 million years ago. There are 35 species, or kinds, of lemurs. All except six species live only on the island of Madagascar which was once called Lemuria. We think of ourselves as very special.

We don't all look alike. Some lemurs are no taller than the length of a banana. Others are three feet high. I am grey, white and black and I have bright orange-brown eyes. My stomach is all white and I like to sit up and face the sun to get warm during the day. My tail is fourteen rings of black and white — that is why I am called a ring-tailed lemur; and while I walk on the ground, I shoot my tail straight up in the air like an antenna. My parents had only one son — me.

Looking back over the years, the first thing I remember in my life was how my mother taught me to eat bamboo. She would break a branch, then tear the bark with her strong, razor-like teeth. I learned quickly to do this for myself. In fact, we ate almost all the plants that grew in Madagascar. We also ate kily seeds, throwing away the rest of the kily fruit and only eating the seed. The place where we lived was beautiful, tall trees bending over unbelievably clear seven small blue limestone lakes. They were called the Lakes of Ifanta.

I clung to my mother as she climbed to the safety of a high branch for the sun during the day or to perch at night for sleep. However, we spent a few hours each day on the ground. The mornings were always fun. We joined the other members of the family in the troop — there were eighteen of us — to search for berries and fruits. The bitter tamarind was a particular treat. And we would play. My cousins were older, but we would meow to make contact, then strut about, our tails straight up, swaying slightly, then pounce, grab and roll together on the ground. We squealed and purred in delight. Our parents paraded about, occasionally sitting to run their tails through their arms. The glands of their forearms gave off a faint scent. They would

*Fosa.*

approach a tree and mark out their territory. Occasionally there were stink-fights when the men lemurs swaggered and threatened each other. This serious game established status and happened in the spring just before they mated. My mother taught me all I knew. She told me about the big island we live on, surrounded by salt water. There are lands beyond, about which she knew nothing, where strange things lived that looked nothing like lemurs. On Madagascar there are no large carnivorates, animals that eat meat. There is only the fosa who looks like a large four-and-one-half foot long black cat. How she knew these things always remained a mystery to me. She told me that Madagascar was not all like where we lived. There were deserts in the south, not too far from the seven lakes of Ifanta, and plains, tropical rain forests and volcanoes in the center of the island. In the north, orchids grew on the ground and clung to trees.

When people first arrived in Madagascar, they came from a faraway country of islands called Indonesia where people grow and eat rice just the way they do in Madagascar. Also, strangely, there is a lemur, the Loris, in Indonesia.

Mother told me many stories about other animals in Madagascar. She spoke of the tall — more than three times the height of my father when he was sitting up — Indris, whose cries pierce the tropical rain forest, scaring people. The idea appealed to me. She also told me of the Daubentonia. That is his Latin name; he is also called an aye-aye. Mother said he looks very strange, with teeth of a rabbit, ears of a bat, fur of a wild boar, hands of a monkey and tail of a fox. He has a twig-like forefinger which he uses to tap on trees, searching for ants, then wiggles his finger into a hole for the delicious larva. She told me I would never meet an aye-aye as their home was far from ours and that there may be as few as fifty alive as people had caught them and eaten them. How horrid! How could they do such a thing, especially as the Malagasy believed that anyone who ate an aye-aye would die within a year.

The lemur macaco rufus and lemur fulvus collaris look somewhat like (though personally I don't think so) lemur cattas. I liked what Mother told me about the lemur macaco or black lemur because she said that he had a ruffled hairdo and is a friendly creature. Maybe someday, somehow, I would meet one, though I doubted it.

*Lemur fulvus collaris.*

At home we frequently saw Malagasy nesomyine rats, whose character was commendable, whose nature was gentle. They were real gentlemen and lady rats. Not like the rattus rattus who came to Madagascar from abroad on ships and who spread diseases. I had met a tenrec, one of a large family. He had fifteen brothers and sisters, which is not unusual for a tenrec. We could understand each other quite easily, but I couldn't understand a thing when he spoke to his brothers and sisters. He would vibrate the quills on his back, making an ultrasonic sound, which his siblings heard but I didn't. We played together. He would wrinkle his forehead up with a big frown. Then he would roll up into a ball and I would push him with my nose. Once he rolled into one of the seven lakes, but it was shallow and he wasn't hurt.

My nervous, busy, curious best friend was a mongoose, the galidia elegans. He didn't come from southern Madagascar and had found himself near our home by chance. He had been looking for food and jumped into a truck way up north. Before he could get off, it was moving and shaking which it did for hours. Finally when he was able to crawl down, he was far from home. I envied him his experience, although I was sad for him because he must miss his home.

This was our life — as I knew it. Sometimes one of us (I tried it once and was punished by mother's caring bite) would wander off toward the Onilahy River where we heard noises and knew there was excitement.

# CHAPTER TWO

One night I woke with a start, nearly losing my grip around my mother's waist. There was a terrible wind and the branches of the trees were bending, water from above falling everywhere and running into the seven lakes. The sky was dark. My eyesight is usually good at night, but I couldn't see at all. I lost my hold and fell from the tree hitting the ground hard, lying there stunned. I'm not sure how long. I was not hurt, but when I cried for my mother, there was no reply.

I lay there huddled against the bark of a tree until the night passed and light came again. I was scared. The troop had left; there was no one. I was alone beside the seven lakes, which always had looked friendly but now gave me no comfort.

I waited awhile until I heard some sound. It came from the Onilahy River and got louder. Two small people appeared at the far side of the water's edge. They were of the Vezo tribe. I had seen the Vezo when they came to take away the bee boxes which hung from the trees at seven lakes. I tried to look very small as they approached. When they got very close, I decided to run. If my mother had been there she would have known better. I

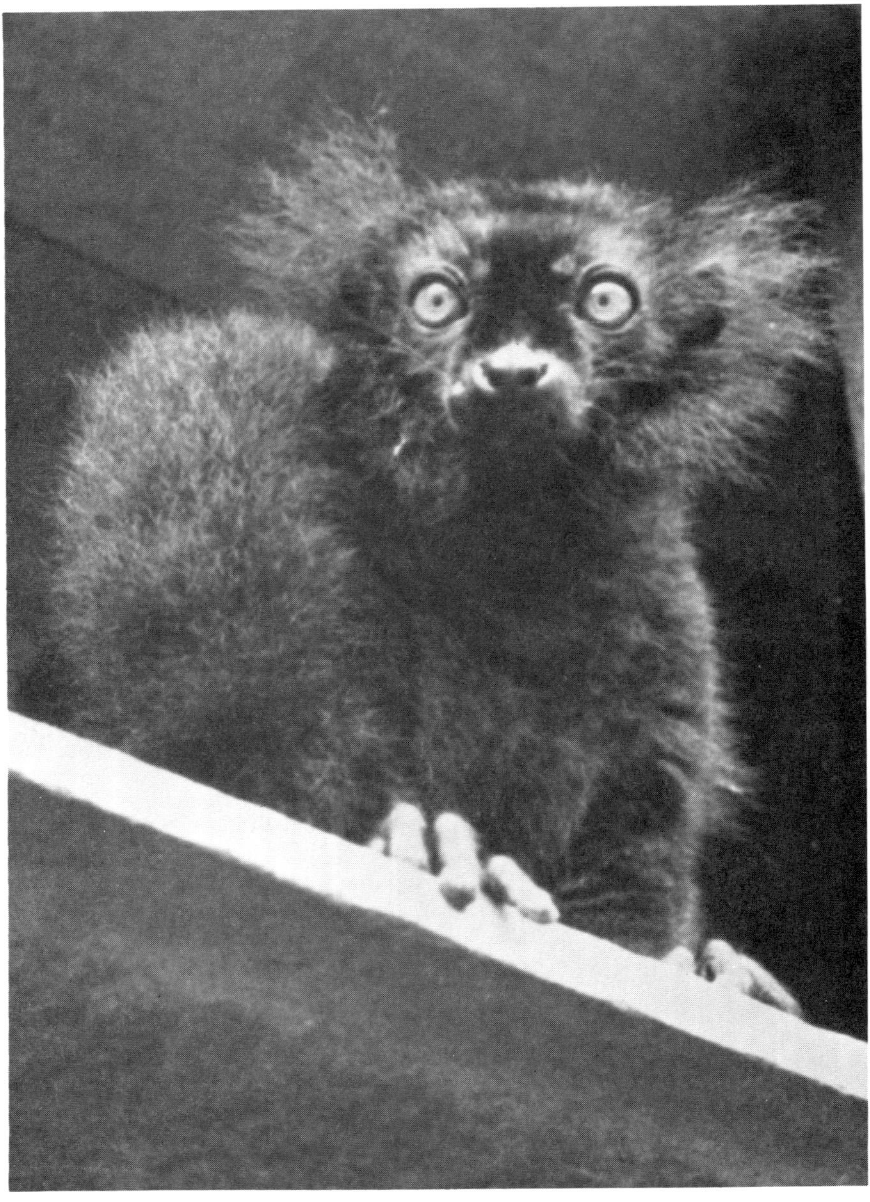

*Lemur macaco macaco (black lemur).*

*Dimy and Basti with the Vezos: "I clung to his shoulder, my tail straight down."*

thought I was free, but the next moment I felt branches all around me, catching me, holding me. It was a net. I struggled hard, the net only tying me up into a knot. My heart beat very fast. I knew they would eat me as people had eaten the aye-aye mother had told me about in northern Madagascar. My friend the tenrec had told me tearfully that cousins of his had been eaten too. I made loud clicks and grunts. They took me to the road and walked down it. I saw what I knew was the Vezo village on a road at the edge of the Onilahy River.

I was held to the ground, and the net was untangled and taken off. Hands put a cord around my waist. I could run, but I couldn't go anywhere. I was tied and put in a dark room that smelled of fish.

Sometime during the early morning a man with a long pole came into the darkness and made a light. He unslung a long fish — almost as long as the man was high — from the pole and put it down on the ground. It was very thin, wet and dead. The man tightened the cloth about his middle, stretched, looked at me briefly, put out the light, and left. It was all dark again, and I thought of home, the seven lakes and my parents. I may have whimpered a bit and made soft clicking sounds which I do when I'm worried. The cord was tight, and hurt. I licked some water from wet leaves on the ground near where the fish lay. My mouth was less dry, but the taste was oily. Even though it was dark, I could see the movement of a number of small bugs — I think they were beetles — across the dirt floor, but they weren't bothering me. I left them alone. I liked the leaves and bark, which they had left for me in a small pile, but I was too worried to eat.

I heard noises outside and guessed I had been asleep. Light found cracks in the wall and I could see clearly. My eyes were beginning to adjust when I heard a very loud noise, like thunder. But it didn't stop; instead it got louder, going bang, bang, bang, bang. It was then that the door opened and a Vezo man in a white straw hat, open white shirt, faded blue trousers and a grin on his face rushed in, scooped me up in his arms and took me outside. The whole village had turned out to see the thing that made thunder. The little girls in bright colored dresses, the mothers, fat with new children, and the young boys wearing shorts and tee-shirts. The thunder stopped and I saw that the noise had come from a large machine with wheels. I learned it was called a landrover. Five white-faced people got out of the machine. I had never seen white people before. They covered their eyes with dark glasses so I couldn't see their character, but they were smiling which reassured me. My captors and the people from the machine talked and kept looking at me. A white man with a broad pleasant face reached out for me and I clung to his shoulder, my tail hanging straight down. He didn't hold me too tight and stroked my neck. his arms were hairy, unlike the black village people. Maybe his parents or great-, great-grandparents had been related to lemurs, I joked with myself as I gained some confidence. It was just then that a white-faced lady with a cloth about her head took me and we — to my utter surprise — got *into* the landrover. I looked out to see one of the white men giving a villager a piece of paper. They nodded, shook hands and said a few more words before he also got into the landrover and we moved off. A black man, but not from the Vezo village, sat behind a wheel and directed the car. The other strangers, all white-faced, were three men and two women, all of whom I got to know well in the days that followed.

# CHAPTER THREE

As I have said, I could not understand people language, but I am a quick learner. They neither clicked nor barked. Even to this day I cannot say that I am bilingual. For one thing I cannot *speak* one word of people language. I found that even people speak in different ways and often have a problem understanding one another. But I did learn to know what they meant from their gestures, tone, and even remembering that a particular sound meant a particular thing.

The first people word I learned was banana. I was on the lap of the lady with the head cloth who sat next to the driver and fed me small pieces of a banana in her fingers. The dust from the road blew all around us making a tan cloud. We were tossed about for a long time until the landrover reached a smooth-surfaced road, lined with palm trees. The palms' fronds rustled from a slight breeze. We went faster until we came to a large town. At least I thought it was a large town. The people called it Tulear. I was tied to a tree while the people sat and drank tea. I was given milk. How did they know I would like milk?

This gave me time to study the five people. There was still

some sun, so I sat upright, facing the sun's warm rays, and examined their faces and the way they looked. First, I had to give them names. I learned their real names much later, but I couldn't continue just calling them the five white-faced people. With a flash I knew what I would call them. I would call each of them by the first five numbers in Malagasy: iray, roa, telo, efatra, dimy. There were three men. They had all taken off their dark glasses so I could now study their character from their eyes. The one with the big hat and pipe, grey hair, tan shirt and short trousers would be Iray. He appeared to be thoughtful, considerate, intelligent and forceful. The man with a tan jacket and long tan trousers, whom everyone greeted when we stopped, I would call Roa. He seemed to like me. I just felt it, the way he rubbed me behind my ear. The lady (I found out later she was Roa's wife) I called Telo. I felt she was a little scared of me. Lemurs have very sensitive noses and can sense these feelings. Efatra was the nice lady with the scarf who held me on her lap and gave me a banana. She treated me like a relation. Dimy was the man who first took me from the Vezo and held me protectively. Well, that was that. Now the people had names. I felt very pleased with myself.

It was dark and I had had the longest, most exciting day of my life. I was all ready to go to sleep. Instead, Iray, Roa, Telo, Efatra, and Dimy crawled into a different automobile, nicer than the landrover. The purr of a quieter engine moved us out of Tulear.

When we stopped Iray gave me a pat and left me in the car, but I could see what was going on. They all went into a house, although Efatra returned with a dish of water, which I spilled all over the floor of the car as I hurriedly drank it. Dawn broke.

*Basti and Iray, who was "thoughtful, considerate, intelligent and forceful."*

They all came out of the house and walked over to another building into which dozens of children had entered and were singing. I gathered from what was said that this was a Lutheran Mission School. Farewells were brief. Dust filled the sky as we sped down the road.

Still driving — where were we going? Would we drive forever? Midmorning we stopped, again with missionaries, but this time it was a place for sick people, a hospital. I could smell the medicines long before we came to the cluster of buildings. It made my nose twitch, and I shook my head.

The place was called Ejeda. These missionaries were very nice. A doctor gave me my first dried raisin. It was chewy, sweet and juicy. He gave me several. I wanted more. The people ate cakes, cookies and yam chips. Iray and Roa experimented to see whether I liked any of these, but the raisins were my favorites. I

decided I would do anything to get more raisins, but there wasn't anything I could do. Our stop here was short.

The landscape had changed dramatically. No more tall green trees I might climb whose leaves provided food and shade from the sun. It was flat, dry, sandy, hot. There were strange trees called "baobabs," known to the Malagasy as "big-bellied trees." Their bases looked all bloated and their branches reached out to the sky like wild roots. African legend has it that the baobab somehow offended God who punished it by placing it bottom-side up. We stopped at the badlands, near Ampanihy. I had never imagined there could be a place so bleak. It was all rock — big ones, small ones. The sun was moving low on the horizon and shed a scary reddish-purple light over the rocks. They seemed to glow like live coals. It looked like the end of the world, a place where time had ended.

# CHAPTER FOUR

Then we motored on to a forest. They called it a forest, and the trees were high, but not my kind of trees. It was a forest of very high cactus called Didierea. They were several times the height of people, the branches covered with thorns reaching out like wild arms. Shorter, scraggly bushes contained a milky sap inside them which would blind you if it got in your eyes. I saw two big snakes slithering through the bushes, but they didn't stop to introduce themselves. Personally, I don't find snakes very friendly, although in Madagascar they are not poisonous. A forest cuckoo with its yellow breast and blue tinted eyes flew by, waving a cordial wing at me, but that was all.

Just before dark we arrived at a broken-down building called a hotel in Ampanihy. A large tamarind tree stood next to it. The people said they didn't like the hotel. I didn't understand as there were very friendly mohair goats, rats, cockroaches and several other bugs living there with whom I had a nice time talking when the people went off for dinner. The rats lived in cement holes inside the house. They told me there was always plenty to eat from the kitchen. Though lemurs eat insects, I was longing for conversation. Cockroaches were everywhere. There were so many of them they said they were critically overcrowded. They feared there might be a housing shortage one day soon. I am not fond of cockroaches, not even for a snack, and I certainly never talked to one before. But the elderly and rather distinguished cockroach who introduced himself to me was different. He was well-mannered, polite, and intelligent. he told me about the Aepyornis, a bird which had lived in this part of Madagascar until 300 years ago when he became extinct. He was much larger, the cockroach said, whisking away a large breadcrumb from his path, then the ostrich of Africa, which he resembled, and weighed six times more. Aepyornis laid huge eggs. The cockroach, who introduced himself as Sam, said that people had found as many as eighty eggs in river beds, some quite intact and well preserved.

"Why would they want a 300-year-old egg," I asked, giving my tail a flip. "They cannot eat them." Sam ignored my question and went on to tell me about tombs, and how the Malagasy people felt strongly that death was a happy thing and that while the eighteen different tribes of Madagascar each had different kinds of tombs, all glorified their ancestors. In the south — where we were — there were tombs that had wooden poles on top of which stood carvings. A painted carving of a car might mean that the man had taken a trip when he was alive. Many tombs had the horns of sacrificed cattle buried in the ground with the points up to show that the man had owned many cattle, that he had been a wealthy man. I asked Sam about the missionaries, telling him about my experiences of the past day. He told me that the missionaries had come from the United

States in 1868 and still considered Minneapolis their home, although they lived, married and stayed in Madagascar. He said that some of them were nice, but that their homes were too clean for cockroaches.

I said, "Sam, how do you know about all these things?" I was amazed. "Well," he replied, reshuffling the same, dried, grey breadcrumb, "we cockroaches have a very good communications network throughout the country. It is that simple."

Sam and I had been talking next to the car, as that is where I had been left, my string tied to a doorhandle. "Do you want to see the inside of the hotel?" Sam inquired. I replied that I thought it would be fun. It only took me a few minutes to work the knot loose with my teeth, and we were off. This was the first house I had ever seen, and I must say I agreed with the people—it really wasn't all that nice. But, of course, I had nothing to compare it to. The beds had nets over them, which, so Sam told me, kept the mosquitoes and rats away from the people at night when they slept. The main room is where I really had fun. A long high table stretched alongside of the room. Sam said it was the bar. Bottles were on the shelves behind the bar. I jumped up on the stretch of wood and slid down its length, knocking over several glasses. This was great. The glasses made tinkling noises as they fell onto the cement floor. It was when I decided to leap to the lines of colored bottles that the people returned and came into the room. They seemed very excited, which I couldn't understand as I was enjoying myself, forgetting family and home. They started to chase me, and the more they chased, the more bottles I knocked over, sending them crashing to the floor, smelling up the place terribly. Finally they did get me, put me inside the car, closed the doors and left me for the night. However, Sam was able to squeeze in through a crack and we talked long into the night. Sam spoke of Madagascar's 800 different kinds of butterflies, half found nowhere else. Also about Madagascar's more than 300 different birds, of which 46 species, or kinds, of birds are found only in Madagascar. He whispered to me, as though it were a government secret, of the coelacanth, a prehistoric six-foot, 150 pound fish with fins it had once used as feet to crawl on land. The coelacanth was thought to have become extinct 20 million years ago. However, it was rediscovered in 1938 in the sea between Madagascar and the Comore Islands.

Sam and I said goodbye just before dawn. He flattened himself before sliding under the frame of the car door. I followed his path from the car window. He scurried across the dirt, entering the house brazenly, skillfully avoiding Dimy's foot. Telo was the last person to get in the car. Iray was rubbing his eyes; it seemed they had spent an uncomfortable night.

## CHAPTER FIVE

We spent a long day on the road, but I found it fascinating. My eyes were opened wide by the new sights. First we drove to a place where both old and young women were spinning mohair without a wheel. The car was parked directly in front of the entrance to the factory. The window had been left open slightly, so I could see, hear and smell all. They made rugs by tying pieces of mohair — the hair of a goat — individually, piece by piece, onto a giant loom. The place smelled heavily of lanolin from the wool. I wondered whether it hurt the goats when they sheared the hair from them. But I saw several long-haired goats wandering around to the side of the building, munching away at scraps and grass. I tried to attract their attention, by jumping up and down on the back seat, but they didn't seem to have a care in the world. I guessed that the cutting of their hair did not hurt them.

We stopped to see more tombs. Different ones. I liked it when we stopped for lunch. Iray and Roa gave me tastes of everything they ate. They — and I — had papaya and banana, chicken, mashed potatoes, eggplant and rolls. The eggplant was somewhat bitter. I didn't like the chicken. The papaya and banana were best, and I licked my fingers long after I had finished eating.

After lunch, as we were moving steadily down a dirt road, there was a sudden bang, a very big noise. Louder than the bang, bang, bang of the landrover. Then the car started to shake. The driver stopped, and everyone got out to look at the back of the car. I rushed to the back window and peered. It was a flat tire. One of the men came back and let me out of the car. I understood it would take some time to fix the tire. I was allowed some freedom at the end of an extended string. Iray had replaced the tight one the villagers had put on me with a more comfortable one. I jumped up and down with great excitement when I saw a tortoise crossing the road directly in front of the car. I rushed over to meet him. He stopped, drew his head into his shell and was still. I circled him. His spherical shell was beautifully patterned and looked polished. As I came around him for the third time, he stuck his nose out, enough to see. When he saw there were no people he came out further. I sniffed at him and asked him his name. "I am a radiated tortoise," he replied, "and I feel sick." With that, he threw up right in front of me. "What is the matter?" I asked, caring. "I ate something old that disagreed with me. Please forgive me. I'm most inhospitable. I'm not always like this. It's an off day for me. It has been extra dry around here and I haven't had enough water to drink. Who are these people?" he pattered on. I told him my story. "Where are you going?" he inquired. I told him I didn't know, that we just seemed to drive on and on, perhaps we would be driving for the rest of my life. I told him the people had

spoken of the sea. "I've never been to the sea," he said sadly, stretching his neck a bit to the left. "I wonder what it is like?" I told the radiated tortoise what my mother had told me about the sea, although even she had never seen it.

The tire had been fixed, a new wheel had been put on, and Roa said that they were ready to move on again. As we got in, I noticed a paper cup with water on the car seat. I knocked it over with a swish of my tail. It fell from the car, magically landing straight up. I hoped the tortoise would find it. As the car started, I stole a look, and saw him slowly moving toward the cup.

We arrived at Faux Cap, which in French means False Cape, by late afternoon. I knew it was called Faux Cap, because they said so, and I understood. I was quite please with myself. A very nice French couple named Balzer met us. They lived and worked there for a French government company that did maritime research. They had two small children. Dimy stayed with a family in a lighthouse, but I stayed with Iray, Roa, Telo and Efatra who were shown to a comfortable house with a magnificent view of the sea. The sea! I could never have imagined that it would have looked like this. Moving blue-black water as far as I could see, with streaks and patches of white. Just below me, a beach — sand from left to right, the sea moving in on it and then back, making a pleasant comforting sound. Above the beach flew clouds of birds, banking one way, then the other, darting low to the waterline, then soaring up into the sky. Free and happy. The sun was starting to set on the horizon turning the sky and sea into a mixture of light and dark colors. I was amazed and stood still, which for me is strange. The people walked on the beach, and I went with them. My string was long and allowed me to chase small bugs that disappeared into the sand and explore shells, some alive with an occupant. I gripped some seaweed in my hand, sniffed at it, then tasted its salty surface, and threw it away. I listened to Roa and Iray talk as there were no friends for me. I heard Roa say that, except for a few small islands, there was no land between us and the South Pole and a continent of ice. Now that I had seen the sea, I didn't think I would be too surprised by a continent of ice. Nothing could surprise me after this!

As night enveloped us, the stars shone in the sky as I had never seen them at the seven lakes of Ifanta. Especially bright were the stars forming the Southern Cross. And then the moon rose from the sea like a large round glowing fruit, sending a line of orange across the water. As it rose, we returned to the house of the Balzers for dinner. I was carefully tied to a solid piece of furniture in the corner of the room, while the people sat at the table and ate. I twisted about a bit, but settled down when they gave me a banana to eat. I peeled it carefully, enjoying myself, and watched them stuff themselves with seafood. First, dishes of enormous oysters were set on the table. The oysters disappeared and the shells were piled up. Telo and Dimy said they had never eaten oysters before, but they stuffed themselves silly with the wet slimy things, and didn't even offer me one. Then, small oysters were brought to the table, and they disappeared, the people making many satisfied noises. I wondered why, but didn't find out. Next, trays of small grilled lobsters, and finally a very large, meaty, and tasty — so they said — fish called a Lamatra. Topping all of this, a souffle. "Magnificent," the guests all cried out, complimenting their hosts. I slept soundly that night to the new sound that had been introduced to my life — the calming sound of waves.

The morning came, and the color of the sea had changed. As our journey continued, I was sad to leave this new experience. Midmorning found us at a marketplace. I had never seen so many people. They were everywhere. Some were leading cattle, their big horns looking heavy on their heads, to places of sale. Goats strayed aimlessly through the crowd. Stalls were set up in groups and people clustered about, buying and selling. The men wore straw hats that looked like beehives and carried sticks which they used to prod the cattle or to lean on while deciding what to do next. Cloth and clothing were both hung from ropes between poles and also laid out neatly in piles on the ground. The cloths were colorful, and many had patterns or pictures.

Before noon we arrived at a sisal plantation. There were

large stretches of land where the sisal grew, a long sharp-pointed, cactus-looking plant. Further on were a field of wood crosses, on which the picked sisal drooped, drying in the sun. I heard great activity in the factory where the sisal was placed in vats and passed through machinery to make it into string, rope, mats and sacks. All this interested Roa very much. They asked many questions from the owner and their host, a smart-looking Frenchman named Jean de Heaulme, who had come to Madagascar in 1928 from an island near Madagascar called Reunion. The people ate again. They always seemed to be eating. Of course, I liked to eat too. And, I did. They had something called osso-bucco that didn't appeal to me. But I liked the stuffed eggs, cheese, cake and fruit.

It was after lunch that I got a big surprise. We all went off in the car to explore Mr. de Heaulme's forest reserve, a protected piece of land — for, guess what?, lemurs! I was thrilled when I saw them from the distance. They were high in the trees, some shifting from one tree to another. At first I couldn't see them clearly, then I heard them calling to each other. They were lemur catta. I made lots of noise myself, meowing for contact and tearing at the wall of the car, rubbing my nose against the glass. I threw myself about with great gusto, making Telo quite unhappy. But she understood why I was acting this way. Lemur catta! But we didn't stay, and soon I saw a different lemur, the sifaka. He looked very handsome, so well groomed, and so big. But introductions were not in order. Could I escape? I realized this was impossible. I mooched, disappointed. I sulked in the corner of the back seat, thinking of home, of mother, of the seven lakes. I hoped my family was all right.

# CHAPTER SIX

The car stopped. The engine went dead. The people were quiet and told me to be. I already was. They put binoculars to their eyes as Mr. de Heaulme got out of the car and clapped his hands. In a moment the branches in the trees became alive and the open sky above was blackened by hundreds of reddish-furred fruit bats. Their five-foot wings made a rushing noise. After minutes they started to settle down again, reaching a branch in flight, grabbing on to it, then hanging, head down. The branches, high above us, became heavy again with these foxy-looking creatures.

We did not stay with the nice Mr. de Heaulme for the night, but pushed on. I curled up and I think I slept a bit, tired by everything. I was startled into consciousness by the car door closing, and shot up to look out. It was dark, another day in my new life of adventure having passed. There were fir trees and palms and houses, many of them, not at all like the houses of the villagers on the Onilahy River, or even the houses we had stopped at on our trip. These houses were sturdy, and there was short grass all around them. Flowering bushes growing at the corners of the houses giving them cheer. I didn't have much fun this evening as I was left alone, after a bowl of soup, and there was no one to talk to. I was just a little scared. It was quiet out, although I thought I heard the ocean. Perhaps it was my imagination.

I thought we would not see Mr. de Heaulme again, but in the morning he appeared. I associated him with the lemurs he protected on his sisal plantation. I liked him and showed my happiness in seeing him by licking his hand when he reached for me. He took us in his car on a tour outside the town of Fort Dauphin where we had spent the night. We were near the sea. We passed beautiful beaches, not the same as those at Faux Cap. More beautiful, in a different way. They were curved by the land, and bordered by different kinds of palms and bright colored bushes. We stopped to admire the travelers palm, shaped like a giant fan. It was called the travelers palm as water was contained at its center which could save a weary traveler in a dry place. But there was plenty of water here — small lakes reflected the vegetation and even some low hills. I caught a quick glimpse of crocodiles, slithering into the water.

The next day something happened to me that I thought was impossible. I must tell you all about it. We drove to a large building. In front of it the ground was cleared and smoothed for a great distance, and there were large machines shaped like birds (I soon found out they were called airplanes), standing on round rubber feet — like those of the car. At first I thought it was a different kind of car, and that the cleared area was a big road. Just as I was getting a good look, I was stuffed into a basket and the top was put on, sealing me in. However, there was a crack in the cover and I pressed my eye against it. I bounced inside my transport as I was carried out of the car, through the big building and onto the airplane. A few minutes after everybody had entered the airplane with the baggage the motors started, making more noise than even the landrover. I still thought we were in a big car as we drove down the clearing and stopped. Then, the engines made a terrible noise and we went faster, faster. From where I was next to a window I could see the land disappearing below us. We were flying. I was in a motor bird. I tried to push my nose through the top of the bag, but Efatra's hand reached in and scratched my neck. It was meant to comfort me. I felt comforted. If Efatra thought she was safe, then I must be safe. I watched the earth pass below us. The houses seemed small. I saw no animals at all from this height. Would I be leaving the land of animals forever?

Was I going to lemur heaven? I wanted to get out and explore, but some voice inside me told me that this would be a bad move. I somehow knew that I was in the bag for my protection, although I didn't understand why. I found that riding in an airplane is different from a landrover or car. The motor of the airplane sounded more like a giant lemur purr. I rather enjoyed it. It was soothing and soon I slept a little.

I felt pressure in my ears. That woke me up. I poked my paw into my ear and scratched, but it didn't help. Nuzzling my way through the top corner of the bag, I stole a look out the window and saw the ground coming up at the plane. We were not going to lemur heaven after all. We were going to hit the ground and the airplane would break up into small pieces. So would we. I started clicking and swished my tail, brushing it

*Antananarivo. "The red clay houses scattered throughout the valleys."*

against the inside of the bag. I was nervous. I was terrified. The land below was different from anything I had seen. There were high mountains covered with trees and rocks. In the valleys were squares of water. I later learned this was where the Malagasy grew their rice. The red clay houses, and there were so many of them, scattered throughout the valleys and perched along the slopes of the moutains. This was certainly different from the desert where we had spent the past week, or, home.

We were falling fast and the friendly purr of the engine changed its voice. Grrr, grrr, grrr. We would most certainly crash against a mountain. The airplane turned, then turned again. I swayed inside my container. We almost touched the trees. And then, the airplane became a car. It was on the ground. The engine became quiet.

We were in Antananarivo, the capital of Madagascar. My new home.

## CHAPTER SEVEN

The year that followed was a happy one for me, although I had a major problem. I grew up and became an adult lemur, and in doing so my character changed. I found that I became temperamental. I didn't mean to get mad, but I did. I threw fits of temper, rushing about wildly, jumping, grabbing at what I could, showing my teeth. I surprised and embarrassed myself when I scratched with my sharp fingernails and tore at the hand that fed me, drawing blood. But that is getting ahead of my story. Let me go back and tell you what happened.

It seems a long time ago since my arrival in Antananarivo, the capital of the Malagasy Republic. Antananarivo is much too long a name for people to say, so they abbreviate it to Tana. Tana is 4,000 feet up from the sea, so it was much cooler than I was accustomed to. Also, it rained a great deal during certain times of the year. Once it even hailed. Big balls of ice fell about me. Fortunately I was under cover, but the noise made me very nervous. I rolled up in a corner, curled my tail over my shoulder, and watched the hailstones bounce off the ground. When it stopped the earth was covered with a blanket of white. Because of the rain and the sun that followed, flowers, grass and trees grew easily in Tana.

But let me come back to all that later. First I will tell you what happened when I arrived in Tana. I lodged with Iray and Efatra at their home where I had a big box. It had holes in it so I could see and get air, but I didn't like it. I wanted to run and play. However, I was only there a short time and it is now vague in my memory.

My new home was my own private house which had been built especially for me in a garden on a hill. I could see the Queen's Palace in the distance. Across the garden was Roa and Telo's house. Between their house and mine was a clipped lawn bordered by flowering bushes. Next to my house were tall eucalyptus trees. I loved the strong smell of their leaves. Beneath the eucalyptus trees Roa and Telo had built a rock and cactus garden.

My house was about ten feet high, six feet wide and ten feet long. It was made of a solid wood frame, a waterproof roof and thick wire netting on the sides, for protection. Roa had placed large branches inside at angles so that I could jump from one to the other, or sit. My favorite sitting place was my terrace, high off the ground. It was there that I enjoyed the sun's warm rays, heating my front and sending comfort throughout my body. At the far end of my terrace was my bedroom, all enclosed except for a door. Every few days Roa would put fresh straw on my bedroom floor, taking out the old. And, I had heat in my bedroom. A lightbulb, shielded so I wouldn't burn myself, was turned on each night. It was cozy inside.

It was on the first day in my new home that I was given my name, Sebastian, and then called Basti (by Roa, Telo and their friends). They had many friends who came to their house.

*Basti relaxing outside his house.*

Sometimes they gave parties in the garden. On the fourth of July they gave a very big party. The garden was filled with people, many of them coming over to pay me a visit. There were other times when Roa would allow me out and I would sit with him and Telo on their large white terrace. My greatest pleasure was when Roa would let me climb the eucalyptus trees. I would rush up to the top of one, spring to a branch, swing to another tree, then perch. I would come down and rush about the lawn, darting in and out of the rock and cactus garden. I would stay still for a moment when Roa gave me a banana, and especially when he gave me raisins. He knew I liked raisins, so he always used them to persuade me to get back into my house, which — much as I like it — always meant an end to my playtime. There were other food treats. The chef in Roa and Telo's house, whose name was Jean Claire, made banana cake especially for me once a week. Oh, what a delicious cake that was — rich, tasty, moist. I ate every crumb. Regular food was lettuce, although I never ate the white part, milk (there was always a dish for me), and bamboo and other green leaves. A fruit I looked forward to was bebas, otherwise known as a medlar (which looks somewhat like a crabapple). I know I was somewhat spoiled by the good food I was given, but I didn't mind at all. I loved every bit of it, and all the attention I got.

I remember once when Roa was away on a trip. I escaped from my wired enclosure when food was being put in. I rushed around the lawn, making a big fuss. Then I saw the water sprinkler. I had seen it before from my house but never up close, so I investigated. I could see that it was ground rain, but didn't realize until it was too late that I would get wet. Which I did.

Soaking wet. And when lemurs get wet, all that fluffy fur gets flattened out and they look skinny and rather sad. Which I did. Telo was very upset. "Basti, Basti, what have you done," she cried. She swooped me up and took me into the house. It was the only time I was ever in it. She dried me off and put me in a room with a machine that glowed and threw off heat. I stayed there until I was dry.

People also came to stay with Roa and Telo. There was one man whom I will never forget. He came in October 1970 and stayed for a week. He had come at Roa's invitation to attend a wildlife conservation conference. I found out all about it as Roa brought the man to introduce him to me and then they stood there, talking. People were meeting from many parts of the world to exchange ideas and try to make decisions that would help save both wildlife and the places where wildlife lived. I was all for that, remembering the nice Mr. de Heaulme.

I heard them speak of the aye-aye and Roa's friend said, "If the aye-aye cannot be saved, then how can Man expect to survive on this planet?" I thought that put the matter very clearly. He obviously cared. Although he went to some meetings, he spent most of his week in a chair in the garden just outside my house writing letters. I found out that he had flown a plane, smaller than the one I had been in, across the Atlantic Ocean. I gathered that was a great distance. In fact, he was the first person to have ever done so. He did it alone. His name, I discovered, was Charles Lindbergh.

One evening he returned from a party at the French Embassy. He and Roa were talking. Roa had a handful of raisins which I was making disappear quickly. I listened as Charles

Lindbergh spoke of an experience he had just had. A Frenchman came up to him at the party and told him he had been at the airport outside Paris when Lindbergh had landed. Lindbergh was interested but not surprised. What did surprise Lindbergh was that this Frenchman admitted that he had been one of those who had torn all the canvas off the plane's structure as a souvenir. Lindbergh said he had never in all the years that had passed met anyone who had admitted to this act of vandalism.

# CHAPTER EIGHT

I had some friends — besides people — but not many. There was a hawk who sat on the rooftop of Roa and Telo's house. But I don't like hawks. They are the one bird that scares me. I duck when I see their shadow on the ground. I therefore never exchanged words with the hawk. However, a chameleon who toured the garden frequently paid me visits, and when I was out for my romp, I would seek him out for conversation and gossip. When they left, I hummed a Malagasy folktune to myself and reflected on the problems people must have.

Leo was my chameleon friend's name. Chameleon, which is a word from the French, means "dwarf lion," which I guess is what the French thought some chameleons looked like. Leo didn't look anything like a lion, at least I didn't think so. But, of course, I had never seen a lion. I knew lions had fur. Leo certainly didn't have any fur. He did not look like a lion. He was about six or seven inches long, liked to walk along the branches of trees or bushes, and changed his color according to his background. Sometimes I would ask him to turn from green to bright red, just for the fun of seeing him do it. When he walked

*Leo (the chameleon), Basti's friend, with Charles Lindbergh.*

he swayed back and forth on his feet several times before stepping forward, with first one foot and then the other. He rolled his eyes around in their sockets, each eye working independently. As I watched him, he would shoot one eye to the left, while the other went to the right. It made it very difficult to tell whether he was listening to what I was saying. I was saying something important. I was telling him about Sam and how Sam knew everything. I wanted to know more about Madagascar. I thought Leo would know, and I was right.

"I always keep a sharp eye out for news," he said, waiting to see whether I would laugh. I smiled. "You already know quite a lot about Madagascar," he said, inching forward on his branch. His tongue snapped out for a bug on a leaf. He was so quick; the bug was there one moment, gone the next. I had already told him about Lindbergh. "What are the Malagasy people doing to save their country?" I asked.

"They are trying, but it is difficult," he replied, looking down then quickly up with his left eye. His right eye was staring straight forward, as though looking for a solution. "It is called the Big Red Island because all the trees have and are being cut down, the land erodes and the rich red soil runs into the rivers and out into the ocean. Also, the Malagasy believe that it is good to burn the grass every year. You lemurs are the real losers. As the trees are cut and the land is burned, you will soon have no forests to live in, no food to find."

"Have they really cut down all the trees?" I asked, deeply troubled. I looked about, across the valleys. There were trees. Leo noticed.

"Of course there are trees, many of them, still on Madagascar. But it is the trend that worries me," he added with consideration. "Some Malagasy care, but everyone has to care. That is why conferences like the one Charles Lindbergh came to are a good idea. It makes people think. Thinking may make people do something." He sounded doubtful. "Time will tell."

*Tsimbazaza in Antananarivo (Tana) the capital of the Malagasy Republic.*

He had edged high on the branch of the podocarpus bush and waved goodbye, saying we would talk more tomorrow.

We did. We had many conversations. In one of them I told him I was getting a bit lonely. At first it was such a new experience for me I didn't mind being alone. Besides, Roa and Telo took such good care of me, and there was that banana cake! But I missed not seeing any other lemurs. Leo told me about Tsimbazaza, the park in Tana, on the other side of the hill where the Queen's Palace stood. Oh, I forgot to tell you, there is

no Queen in Madagascar. There was nearly a century ago, but they still call it that. It's a museum now, and they have a President since they became politically independent from France in 1960. I thought you should know. Leo told me that there was a large zoo and botanical garden in Tsimbazaza. There were lemurs on the other side of the hill and I couldn't even hear them! I let out an especially loud meow and listened. No one had heard me — at least no lemur. I jumped up to my terrace, and after sunning a bit, curled up in my bedroom and slept. I dreamed of seven lakes.

# CHAPTER NINE

I told you that I became temperamental, a little mad, and I didn't know why. That was in May 1971, after I had been living with Roa and Telo for nearly a year. I asked Leo what he thought, but he didn't have an answer. He just looked away with both eyes. He said he didn't know much about lemurs.

But one day I blew up. Roa had come to my house to see me. He spoke to me, as he always did, in French and English. I understood him very well by now. I saw the raisins in his hand. He said kind things, and put his hand in to scratch me behind my ear, which I always liked. I couldn't help myself. I ran my razor-like teeth down his hand, which got bright red and wet. I felt sorry, but instead of stopping, I reached out and scratched him hard with my sharp nails, drawing more blood. I must be sick, very sick, to do such a thing. Roa was shocked, but not mad. He said, "Basti, Basti, Basti." That was all. He went back to his house. I knew I had hurt his feelings.

The next day was just as bad. When Roa came to say good morning, I jumped up to the wire of my house, drew my head down lower than my shoulders, and shook the wire with all my might. I bent my mouth back, showed my teeth, and hissed. I had lost control. There was some chemical change in me. I had become an adult. I wanted a mate.

For the next few days Roa only came to see that I had clean straw and food. No raisins. He kept clear of me, and I wasn't let out of my house. Leo came within earshot the fourth day, but when he saw my temper he disappeared, fast even for a chameleon. Why was I acting this way? Even to Leo.

A few days passed. I lost my appetite and sulked. I had just awakened and had finished my grooming — licking and tooth-scraping my back and belly fur. I had licked my palms and thighs. I was cleaning my ear with my toe, and purring softly when I saw Roa walking across the garden with another man and a box. I threw myself at the wire and hung there. I started to shake the wire but then stopped and watched, curious, for I have always been ever so curious.

They both came into my house, carefully. They took me before I knew it. I wasn't prepared. They put me in the box and closed it. They said nothing.

I was in a car. Were they taking me back to seven lakes? I wasn't sure I wanted to go back to seven lakes; it had been too long. In spite of the way I had been acting, I knew Roa would see that I was not harmed. Maybe. No, he loved me. Still? Yes.

When the box was opened, I found myself in a cage, not a nice house like the one I had had at Roa and Telo's, nor as large. It took me only a second to know where I was. I was in the zoo at Tsimbazaza. There was no doubt about it. There were other cages. There were rows of cages. There were other lemurs, lemurs I had never dreamed I would meet. and, to my utter

*Lemur macaco macaco (black lemur). He had a ruffled hairdo.*

astonishment, in the cage next to me was an aye-aye. There was macaco macaco on my other side, and rufus and collaris down the way but no tenrec.

I still felt a temper inside of me, but the thrill of seeing other lemurs overwhelmed me. However, I could see no lemur catta. No ring-tails, like me.

Fortunately I didn't sleep well that night; otherwise I wouldn't have heard them. I had gotten acquainted with the aye-aye. Mother had described him very well to me; that was

*Lemur Fulvus Rufus*

how I recognized him so quickly. He wasn't handsome, but I had a special feeling for him. He was a sort of star, a hero, for me. He stood for the cause of conservation, of survival. He was a living symbol. He told me of his home on an island called St. Marie, off the northeastern coast of Madagascar. He told me there were only fifteen aye-aye on St. Marie. He had been brought to Tsimbazaza on consultation and then would be returning to St. Marie. He said that while he received an excellent assortment of high quality caterpillar, grub and maggot larva, he missed the excitement of searching for the larva himself. He felt he was sliding into laziness.

*Daubentonia (Aye-aye) at Tsimbazaza. Mother told me I would never meet an aye-aye. She was wrong.*

"Did you know that the Malagasy made a stamp in 1970 and put me on it?" he asked. Of course I didn't. "It was to celebrate the Conservation Conference in October," the aye-aye explained. "Oh, *that* conference," I exclaimed, with a worldly, knowing sound to my voice. "Yes, I know all about it," I half lied.

"It was quite a good likeness of me," he added, ignoring my statement. "An artist came here and sat for hours with a pad. The stamp even had my Latin name, *Daubentonia*, on it. That particularly pleased me, as my mother calls me Tony." He was hanging from the top of his cage, looking down at me as he spoke. It was nighttime, way beyond my bedtime, but not Tony's. He was nocturnal. I am diurnal. He is awake all night, and I do the opposite. He wanted to talk more, but I am afraid that I dozed off for a while. Every so often I would wake up, wondering where I was, then in the moonlight saw Tony wandering about his cage.

It was then that I heard them. "Meow." I was sure. Almost sure. It was a lemur catta. I was very quiet and stretched my neck forward hoping I would hear better. Tony was making some noise against the leaves in his cage, and I asked him politely to be quiet.

"Meow, meow." Then, a different "meow." More than one. I hadn't heard this call for nearly a year. I called back. "Meow, meow, meow, meow" and followed with a few barks and a long howl just for good measure. A chorus of muffled meows sung to me from the night.

I slept very badly the rest of the night, if at all. As soon as there was light, I started looking. Although my eyesight is not particularly good, as with most lemurs, I hoped desperately I would see the lemur cattas. I asked Tony whether he had seen them, but he said "no." He apologized, explaining that he was really somewhat of a loner and hadn't mixed with the group at the zoo.

I spent the day restlessly. I exchanged a few words with the macaco macaco — in fact, there were a pair of them — in another cage. But my thoughts were elsewhere. I remembered my family, my youth. Flashes of the past year sprung into my mind. But what I was really thinking about were those other lemurs.

# CHAPTER TEN

The following day was the best day in my whole life. The head of the zoo, a very kind Frenchman named Roland Albignac, took me gently out of my cage, put me in a car, and drove me around the corner, down a tree-shadowed path to a lake. There were several islands on the lake. One was covered with ibis, egrette, and other herons, flapping their wings, nesting and flying freely in and out of the park. But my attention was immediately drawn to the other island.

Roland Albignac and I got into a small rowboat and headed for the island. My eyes were riveted on what was to become my permanent home. On the island was a troop of lemurs. No cages, no wire. They climbed the tall trees, jumping from branch to branch. Several were sunning themselves on a patch of grass. Others wandered together, their tails straight up. "Oh, glory!" I murmured to myself, purring with delight.

I saw her the moment Roland Albignac placed me on the island. Her name was Colette. She had beautiful dark markings about her eyes. Her belly fur was whiter than any I had ever seen. Her tail had a particular twist to it. She was the most beautiful creature I had even seen. I immediately fell in love, madly in lemur love with her.

We were married at the base of the tamarind tree a few weeks later. It wasn't a big affair. Of course, the whole lemur catta troop was there. A few birds flew in for the celebration. That was all. It was my day, our day. Colette and I were very, very happy.

As this story is recorded forever so that you can read it, I have been on the island for over a year and have become a senior member of the troop. They appreciate and respect my worldliness. I tell the young ones of my early travels. Perhaps I exaggerate sometimes, but not much. Several of the troop were born at the zoo in Tsimbazaza, as were our two children, whom we named Sam and Leo. Colette is expecting another child soon. If it is a boy, we'll call him Roa; if it is a girl, Telo.

FRONT COVER PHOTOGRAPH BY
CHARLES E. WILLIAMS, JR.

ALL OTHER PHOTOGRAPHS BY
ANTHONY D. MARSHALL

# MADAGASCAR

## *The Malagasy Republic*

Madagascar is the fourth largest island in the world, following Greenland, New Guinea, and Borneo. It is 1,000 miles long and 350 miles wide, covering 227,678 square miles. It has 3,200 miles of coast and its highest point is 9,449 feet. More than 6,000 species of flowering plants, 233 species of reptiles, and 142 species of frogs are unique to Madagascar. Principal agricultural products are coffee and vanilla.

No prehistoric man has been found in Madagascar. It is generally accepted that the first people to reach Madagascar were Malayo-Polynesians, arriving about 200 A.D. on rafts. They headed for the hills. The first European to touch on Malagasy soil was Diego Dias, on August 10, 1500. Malagasy kingdoms were formed during the sixteenth and seventeenth centuries. The French took Madagascar by force in 1895-96, maintaining control until the country became independent on June 26, 1960. The first American Consul to Madagascar was John Finkelmeier in 1866.

The Malagasy Republic has a bicameral legislative system. The country is divided into six administrative provinces: Tulear, Diego-Suarez, Fianarantsoa, Tamatave, Majunga and Tananarive. The capital is Antananarivo (it was formerly called Tananarive).

In 1972 a coup overthrew President Tsiranana. President Didier Ratsiraka has been President of the Malagasy Republic since 1975.

**About the Author**

Anthony D. Marshall served with the U. S. Government for twelve years (until 1959, when he was Consul in Istanbul, Turkey). From 1942-46 he was on active duty with the U. S. Marine Corps, attaining the rank of Captain, and seeing active duty with the Third Marine Division on Guam and Iwo Jima. Mr. Marshall was President of African Research and Development Company and Chairman of the Board of NIDOCO Ltd., a Nigerian company, from 1959 to 1969. For eight years (1969-77) Mr. Marshall served as American Ambassador: to the Malagasy Republic, Trinidad and Tobago, Kenya and the Seychelles. He was also the U. S.'s Permanent Representative to UNEP. Returning to the U.S. in June 1977, he has been engaged in business as an investment advisor and consultant on international affairs, with particular emphasis on Africa. Mr. Marshall is an Associate Director of Tucker Anthony and R. L. Day (since 1962). Mr. Marshall is Vice President of The Vincent Astor Foundation, former Chairman of the Seamen's Church Institute, and Chairman of the Board of the Marine Corps Command and Staff College Foundation. He serves on the board of a number of other foundations and organizations, including the New York Zoological Society, WNET-Channel 13, the Metropolitan Museum of Art, Lincoln Center for the Performing Arts, African Medical Research Foundation, Astor Home for Children, Brown University, Board of Visitors of the Primate Center at Duke University, and Rockefeller University Council.

Other books by Mr. Marshall

| | | |
|---|---|---|
| Africa's Living Arts | Franklin Watts, New York | 1970 |
| Malagasy Republic | Franklin Watts, New York | 1972 |
| Trinidad and Tobago | Franklin Watts, New York | 1975 |